GLOBAL WARMING?

Understanding Climate Change by Monitoring the Atmosphere

BY DR. STEPHEN T. HANLEY

The contents of this work, including, but not limited to, the accuracy of events, people, and places depicted; opinions expressed; permission to use previously published materials included; and any advice given or actions advocated are solely the responsibility of the author, who assumes all liability for said work and indemnifies the publisher against any claims stemming from publication of the work.

All Rights Reserved
Copyright © 2020 by Dr. Stephen T. Hanley

No part of this book may be reproduced or transmitted, downloaded, distributed, reverse engineered, or stored in or introduced into any information storage and retrieval system, in any form or by any means, including photocopying and recording, whether electronic or mechanical, now known or hereafter invented without permission in writing from the publisher.

Dorrance Publishing Co
585 Alpha Drive
Pittsburgh, PA 15238
Visit our website at www.dorrancebookstore.com

ISBN: 978-1-6491-3149-2
eISBN: 978-1-6491-3671-8

BASIC CONCEPT AND PROCEDURE

RADIATION SOURCE

ATMOSPHERIC COLUMN

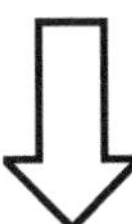

SENSOR PACKAGE

OPTICAL TELESCOPE & SPECTROGRAPH

DATA PROCESSING

COMPUTER & SOFTWARE

VERTICAL PROFILES

MOLECULAR CONCENTRATIONS AND
ATMOSPHERIC TEMPERTAURE

PREFACE

Climate change is arguably one of the most important processes affecting planet Earth, and that will determine the quality of life for present and future generations. It is not enough to simply monitor surface temperatures to understand this phenomenon. We must understand what is happening in the atmosphere.

A method has been developed which allows monitoring the constituents of the atmosphere both in the present and past time periods. The constituents in this monitoring technique include carbon dioxide (CO_2), methane (CH_4), ozone (O_3), and water vapor (H_2O). Each constituent performs functions in the atmosphere, such as ozone (O_3) protecting against harmful radiation from reaching the Earth's surface, and methane (CH_4) and carbon dioxide (CO2) influencing climate temperatures and changes. This monitoring measurement allows more than just a column quantity assessment, but actually reveals how the constituent is distributed along the vertical path, all contained in each measurement. The atmosphere monitoring data may be acquired at regular time intervals (daily, weekly, etc.) to show variances and compared with results obtained from old or archived data sets which were not analyzed with this new procedure. Expensive and complex airborne detector measurements do not provide an instant measurement up through the atmosphere and are not easily repeated on a daily or regular basis.

The following is a description of how measurement of atmospheric transmission can be used to determine molecular profiles or specie concentration along the transmission path. Depending on the molecule under study, methods for monitoring concentrations can be costly and complex in both manpower and equipment, as for example in the cases for CO_2 and CH_4 atmospheric constituents (Peischl 2012; Kang 2012) and H_2O concentrations obtained by combining GPS data and surface meteorological measurements (Bevis 1992). The work described here shows how slant path solar spectroscopy measured at high resolution (<.1cm-1) can be used to retrieve distributions, or concentrations versus altitude, of some molecular constituents of the atmosphere. This work permits retrieval of profiles solely from single-ended ground-based slant-path solar spectra. Results for retrieval of H_2O profiles will be shown here, but other molecules successfully retrieved with this methodology include carbon dioxide (CO_2), ozone (O_3), and methane (CH_4).

Introduction

THE OBJECTIVE OF THIS WORK IS TO extract vertical density distributions of an atmospheric constituent molecule based on absorption detail (<.1cm-1) contained in measured slant-path transmission spectra. Examples of atmospheric transmission spectra measured over long path and the associated hardware are presented in Dowling (1997 and 1998) using a scanning Michelson Fourier Transform interferometer. The measured transmissions must traverse the vertical height of interest for the profile retrieval, as for example with slant path solar spectroscopy providing ground to top of atmosphere or, alternatively, ground to a light source positioned at elevation.

Methodology

The steps used in the retrieval are to perform a high-resolution line-by-line calculation of slant path transmission (McClatchey 1973, Dowling 1977, Dowling 1978) using a mid-latitude profile as a starting point. Next the point-by-point differences between the calculated and measured spectra are noted. Adjustments are made to the initial mid-latitude "estimator profile," and a new calculated slant path spectrum is made. Then point-by-point differences between measured and calculated absorption detail are again examined. When the new point-by-point differences between measured and (most recently adjusted) estimator spectra are reduced, then the most recent adjusted profile (with improvement to fit) becomes the new estimator reference. This process is repeated until no further improvement can be made between calculated and measured transmission curves; at this point a vertical profile has been retrieved. The temperature and pressure dependences of the vibrational and rotational band structure dictate the positioning of the molecule concentrations along the vertical viewing path. Using this methodology, key spectral regions (spectral Keys) have been identified that permit accurate retrieval of water vapor profiles from transmission spectra without interference from other atmospheric absorbing species. The performance of this methodology is demonstrated with two test water vapor profiles. In Figure 2 (Test #1) a simple surface enhancement over what would be considered average concentrations for mid-latitudes, and in Figure 4 (Test #6) a complex set of H_2O layers from ground to 10 km are used to calculate corresponding slant path transmission spectra. These spectra are then treated as experimental or measured spectral "data input," and the methodology is used to extract a best estimate vertical profile for comparison. Results indicate that the methodology works well in both of these cases.

The slant path high resolution transmission calculations were made with a custom line-by-line multi-layer propagation model of the lower atmosphere (0 to 100 km altitude). This model was developed by the author and validated with field measurements (Dowling 1977, 1978). The molecular absorption line widths, strengths, and positions are from the widely available AFGL Line Atlas (McClatchey 1973) which is updated regularly every four to eight years. The model includes good temperature corrections and both collisional and velocity line broadening effects to permit applicability to the upper atmosphere as well as to sea level. The search routine performs trial and error iterations between the measured spectra and an estimator or profile-generated spectra to achieve a best fit. Absorption line shapes are calculated for accuracy at distances near and far from line center over the entire spectrum. If a measured vertical temperature is not available via balloon sonde or other source, then the T-profile can be retrieved from the high-resolution test spectra using this search methodology (same procedure as above) and a different spectral region for the T (temperature) parameter.

The first part of the retrieval problem is to identify a spectral region that provides sensitivity to vertical position of water vapor and is not overlapped or covered by other strong absorbing species nominally present in the atmosphere. Several spectral locations (spectral keys) were identified that provided good sensitivity to water vapor placement with altitude. One of the keys (in the 1.5 micron region) was chosen for these studies and used to evaluate the test spectra. The spectral windows were selected to optimize calculation time and profile discrimination as well as noninterference from other molecular absorbers present in the atmosphere. Two test cases used in this study were generated using two widely differing water vapor vertical profiles to demonstrate the ability of the retrieval process.

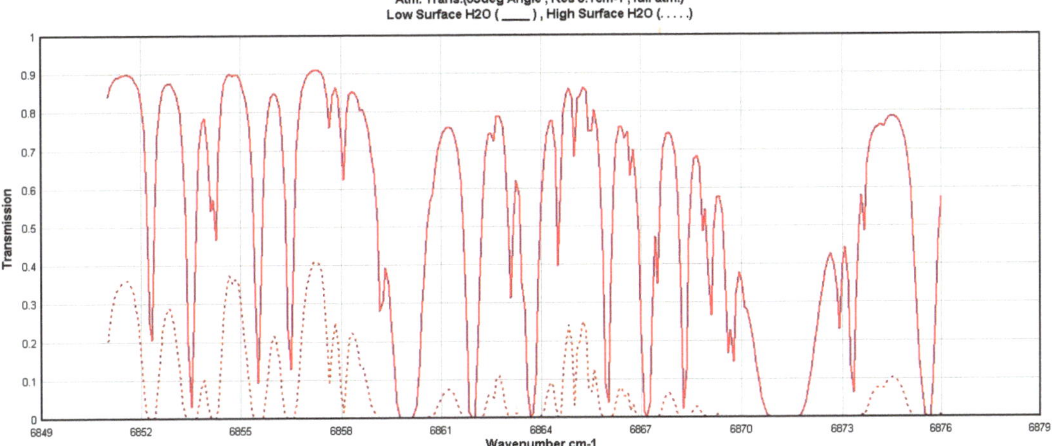

FIGURE 1 shows slant path transmissions over one of the spectral regions used for the H_2O molecule with two differing amounts of H_2O. The spectral differences are illustrated with a dashed curve showing transmission for a high surface H_2O concentration. These differences in transmission detail are used in placing the H_2O in correct amounts (concentration versus altitude) along the viewing or measurement path. Implicit in the differing transmission detail are the temperature and pressure dependences of the vibration and rotational bands within the key spectral region.

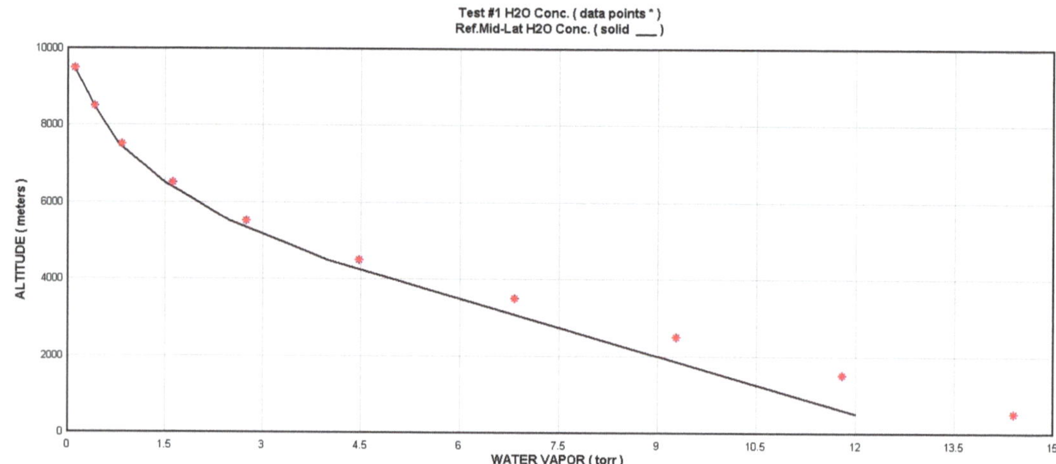

FIGURE 2 shows two water vapor profiles: a typical H_2O profile for mid-latitudes in the solid curve, and larger H_2O concentrations near the surface in the (*) data points. The iterative retrieval methodology (described above) is used to find the unknown profile and the results shown in the following figure 3 below using only the spectral features in the H_2O key region and a first guess mid-latitude profile as a starting point.

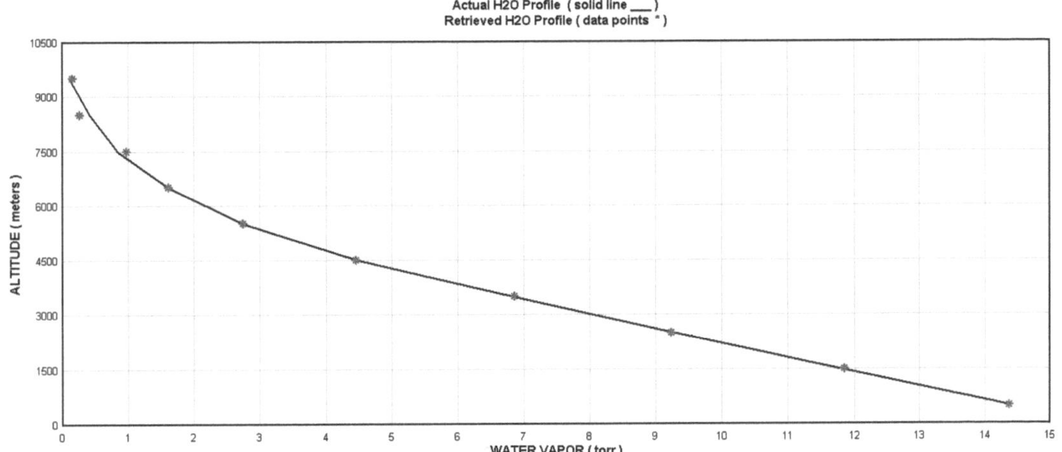

FIGURE 3 shows the actual water vapor profile in this case (solid curve) and the retrieved profile (*) data points based only on the spectrum in the key wavelength region. The accuracy shown here demonstrates the ability of the methodology to retrieve a relatively simple H_2O profile solely from the high-resolution spectrum (Test #1).

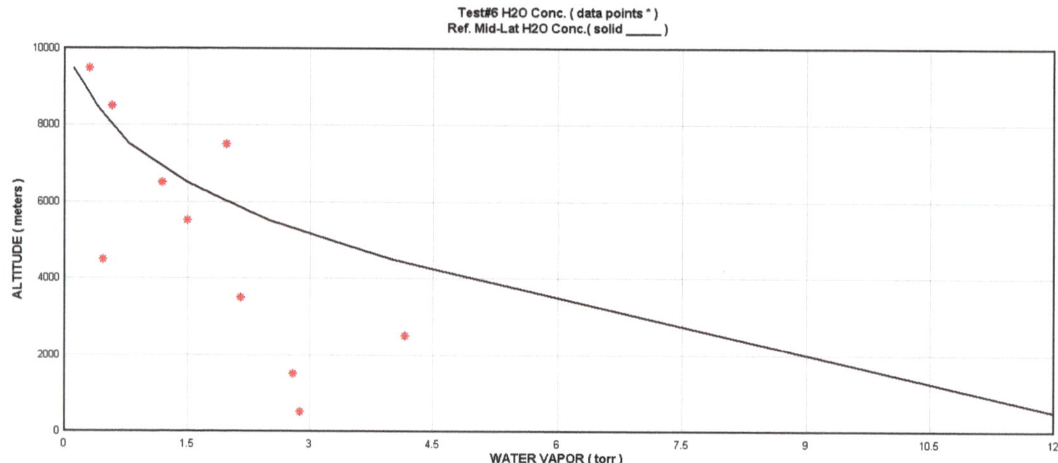

FIGURE 4 shows an average mid-latitude water vapor concentration profile reference (solid curve) for comparison to a more complicated water vapor profile (*) data point. With the mid-latitude profile as a starting point, the retrieval methodology will be used to determine if this more complicated profile can be found solely from the spectral features in the key region.

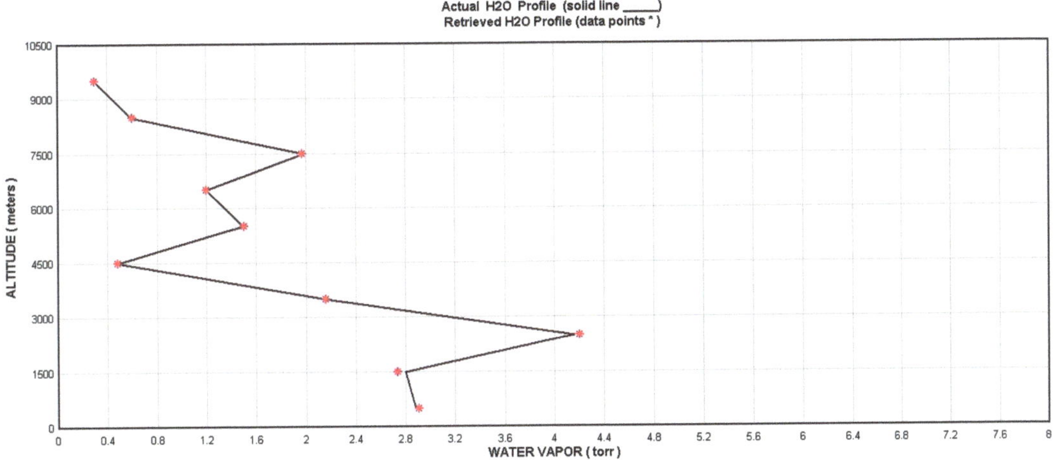

FIGURE 5 shows the actual water vapor test profile in the solid curve and the retrieved water vapor profile (*) data points using an assumed mid-latitude water vapor profile initial starting point. The ability of this iterative methodology to retrieve a more complicated water vapor profile with good accuracy is demonstrated in this case. All retrievals are based solely on the high-resolution spectral features in the spectral key region.

Summary

In conclusion, it has been shown that by examining high-resolution spectral features in key spectral regions, a close approximation of an otherwise unknown vertical distribution (profile) of water vapor can be derived. The speed and accuracy of attaining a good profile approximation is improved by using an accurate temperature profile which, if not available from direct measurement, can be extracted from the high-resolution spectrum using an appropriate T-profile spectral key and the same search methodology described above. Similar results to those presented in this report for the H_2O molecule have also been achieved for other molecules (CO_2, CH_4, and O_3) using appropriate spectral keys suitable to each. The profile methodology described herein was awarded a full utility patent No: 9,335,258 B2 by the US Patent Office on May 10, 2016.

In order to better understand changes in climate, it is vital that we understand the most relevant constituents of the atmosphere and develop the ability to monitor them in a cost-effective and efficient manner.

See Appendix for examples of retrievals for atmospheric temperature profile, atmospheric pressure profile from ground to top of atmosphere, carbon dioxide, methane, and ozone.

References

Bevis, M., Businger, S., Herring, T.A., Rocken, C., Anthes, R.A. and Ware, R.H. "GPS meteorology: Remote sensing of atmospheric water vapor using the global positioning system." *Journal of Geophysical Research: Atmosphered,* Vol 97 D14 (1992).

Dowling, J.A., Haught, K.M., Horton R.F., Curcio, J.A., Garcia D.H., Gott C.O., Agambar, W.L. "Atmospheric Transmission Field Experiments Using IR Lasers, Fourier Transform Spectroscopy and Gas Filter Correlation Techniques." *NRL Reports of Progress* 4142 (1977).

Dowling, J.A., Horton, R.F., Hanley S.T., Haught, K.M. "High Resolution Field Measurement of Atmospheric Transmission." *SPIE Proceedings* (1978). 142, 25

Kang, J.S., Kalnay, E., Miyoshi, T., Liu, J., and Fung, I. "Estimation of surface carbon fluxes and with an advanced data assimilation methodology." *Journal of Geophysical Research: Atmospheres* Vol 117 D24 (2012).

McClatchey, R.A., Benedict, W.S., Clough, S.A., Burch, D.E., Calfee, R.F., Fox, K., Rothman, L.S., Garing, J.S. Air Force Cambridge Research Laboratories Atmospheric Absorption Line Parameters Compilation AFCRL-TR-0096. (1973).

Peischl, J., Ryerson, T.B., Holloway, J.S., Trainer, M., Andrews, A.E., Atlas, E.L., Blake, D.R., Daube, B.C., Dlugokencky, E.J., Fischer, M.L., Goldstein, A.H., Guha, A., Karl, T., Kofler, J., Kosciuch, E., Misztal, P.K., Perring, A.E., Pollack, I.B., Santoni, G.W., Schwarz, J.P., Spackman, J.R., Wofsy, S.C. and Parrish, D.D. "Airborne Observations of Methane emissions from rice cultivation in the Sacramento Valley of California." *Journal of Geophysical Research: Atmospheres* Vol 117 D24 (2012).

Appendix

The following data plots are examples of retrievals for atmospheric temperature profile, atmospheric pressure profile from ground to top of atmosphere, carbon dioxide, methane, and ozone.

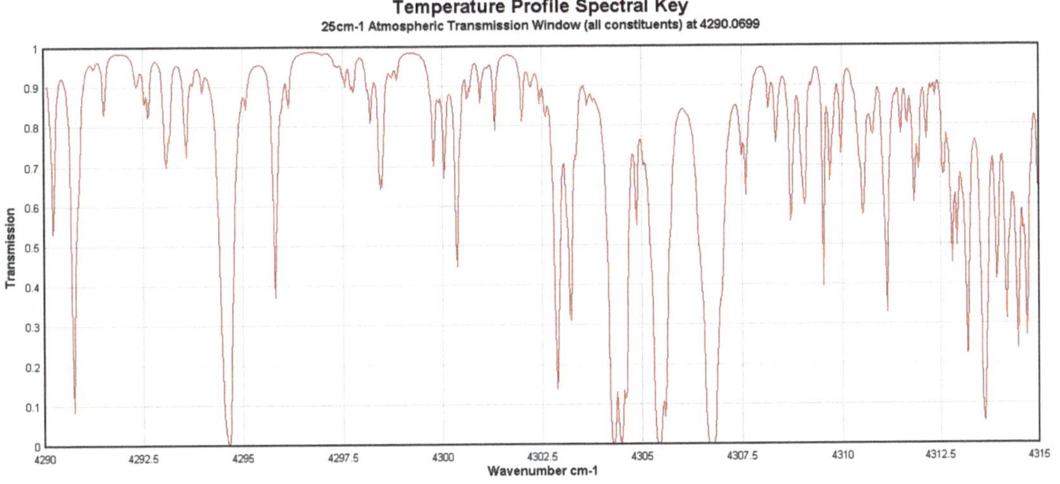

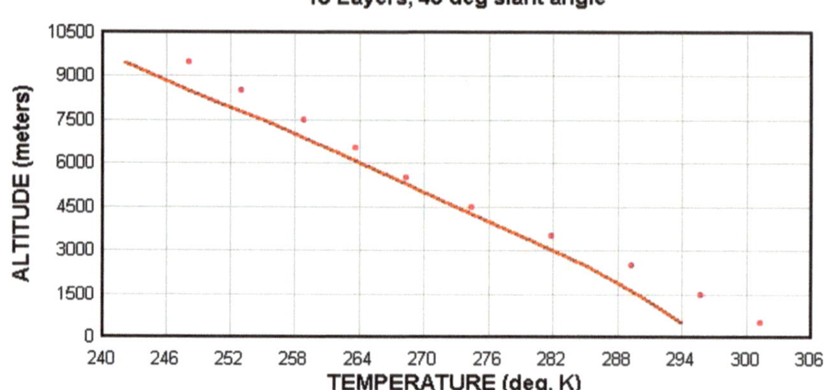

TEMPERATURE RETRIEVAL TEST PROFILE #4
Test Profile #4 (*) , Reference Mid-Latitude Summer T-profile (_____)
10 Layers, 45 deg slant angle
ALTITUDE (meters)
10500
9000
7500
6000
4500
3000
1500
0
240 246 252 258 264 270 276 282 288 294 300 306
TEMPERATURE (deg. K)

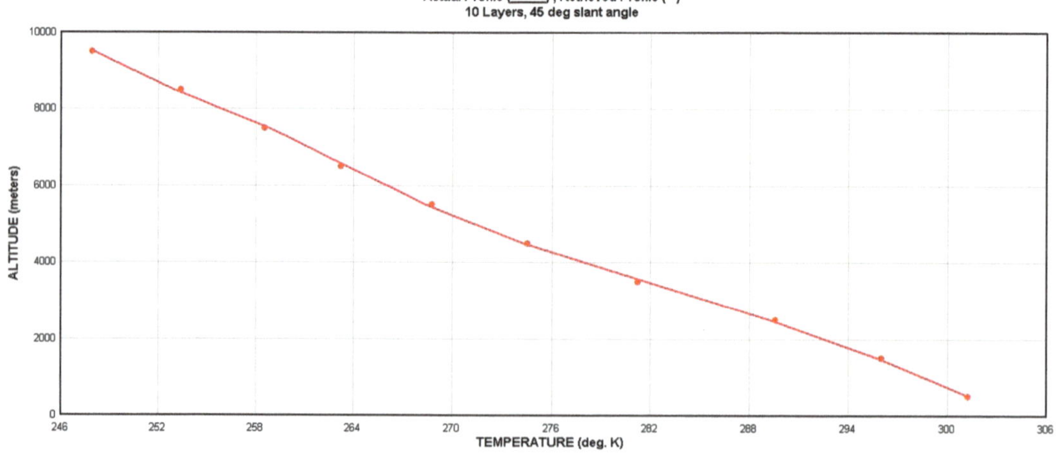

RETRIEVED VERTICAL TEMPERATURE PROFILE (Test #4)
Actual Profile (_____) , Retrieved Profile (*)
10 Layers, 45 deg slant angle
ALTITUDE (meters)
10000
8000
6000
4000
2000
0
246 252 258 264 270 276 282 288 294 300 306
TEMPERATURE (deg. K)

Barometric Pressure Spectral Key
Atmospheric Transmission at 6356cm-1 (45deg slant angle, full atm.)
Low Surface BP (_____) , High Surface BP (......)

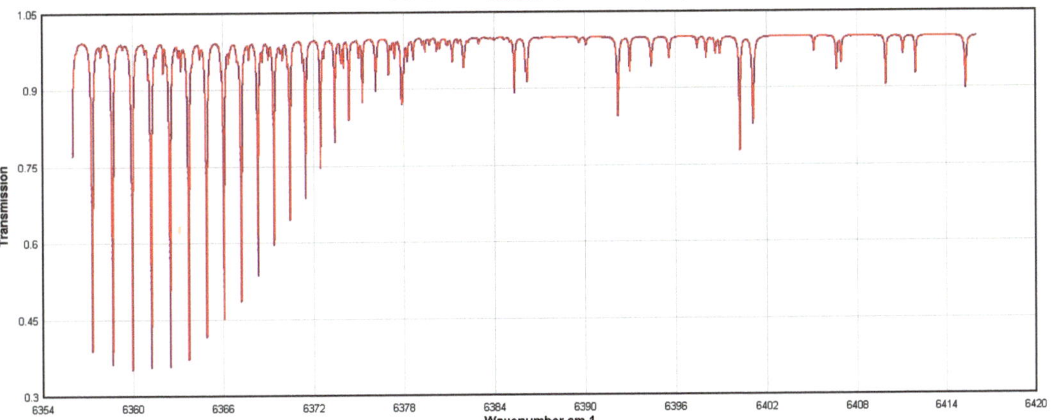

RETRIEVED VERTICAL ATMOSPHERIC PRESSURE PROFILE (Test #3)
Actual Profile (_____) , Retrieved Profile (˚)
10 Layers, 45 deg slant angle

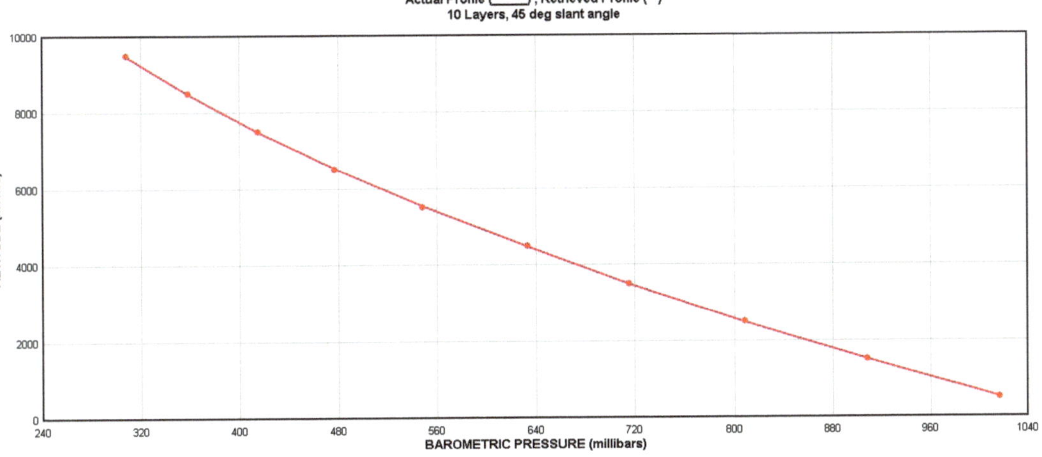

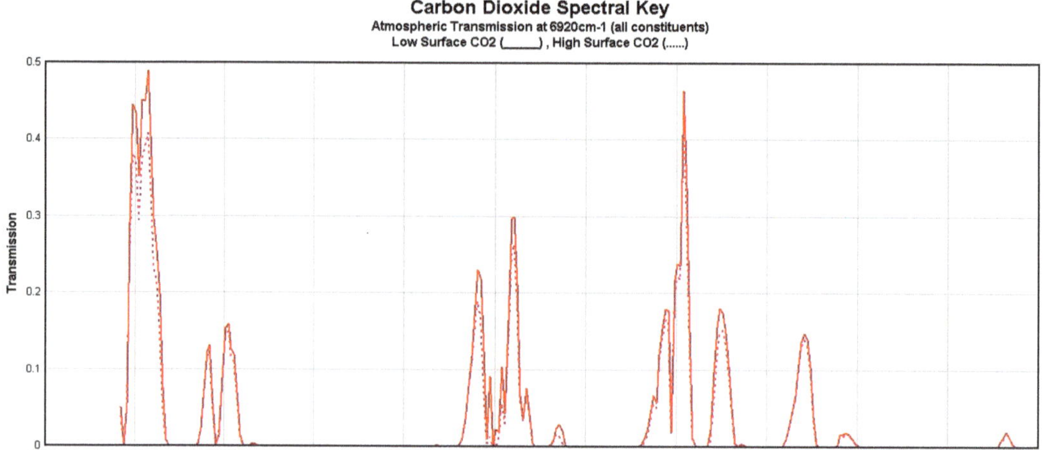

Carbon Dioxide Spectral Key
Atmospheric Transmission at 6920cm-1 (all constituents)
Low Surface CO2 (______) , High Surface CO2 (......)
Transmission
0.5
0.4
0.3
0.2
0.1
0
6947.5
6950
6952.5
6955
6957.5
6960
6962.5
6965
6967.5
6970
6972.5
6975
Wavenumber cm-1

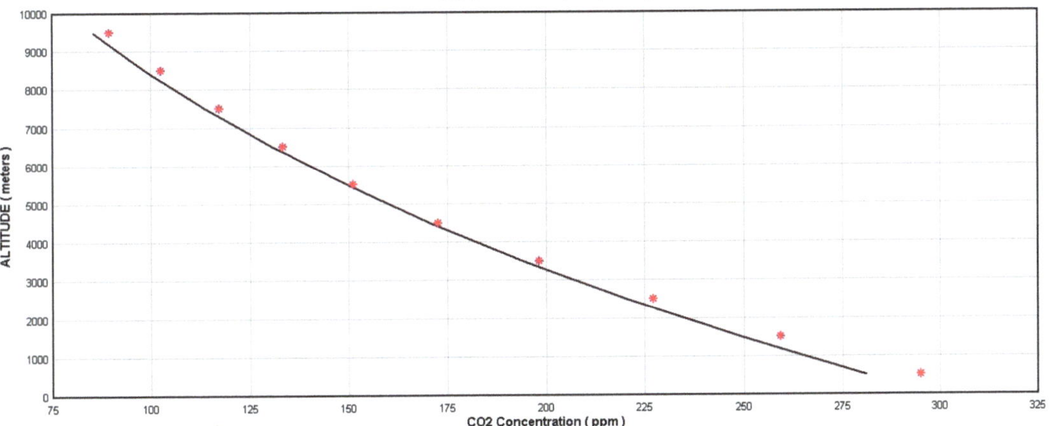

CO2 PROFILE #4, 10 Layers, 45deg Look Angle
Reference Mid-Latitude CO2 Profile (solid line ___)
CO2 Concentrations (data points *)
ALTITUDE (meters)
CO2 Concentration (ppm)

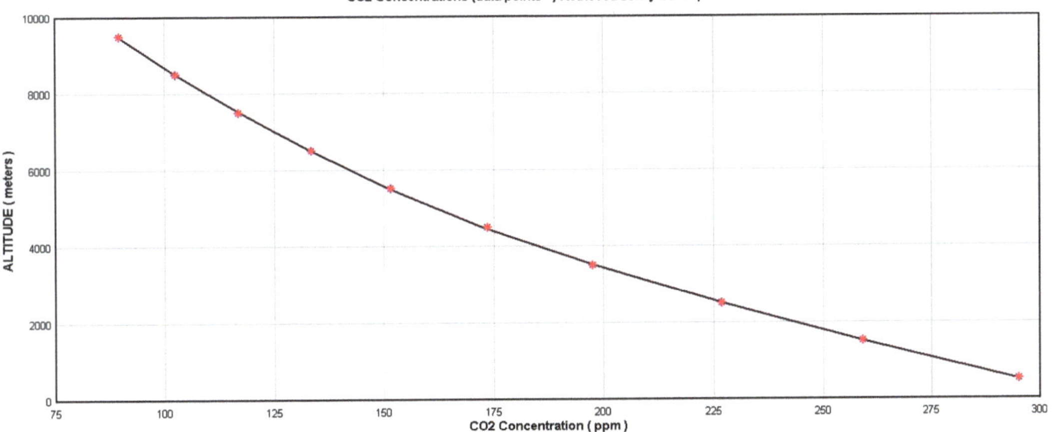

CO2 PROFILE #4, 10 Layers, 45deg Look Angle
Actual CO2 Profile (solid line _____) , Spectrum Res. 6.8587cm-1
CO2 Concentrations (data points *) Retrieved solely from spectrum
ALTITUDE (meters)
CO2 Concentration (ppm)

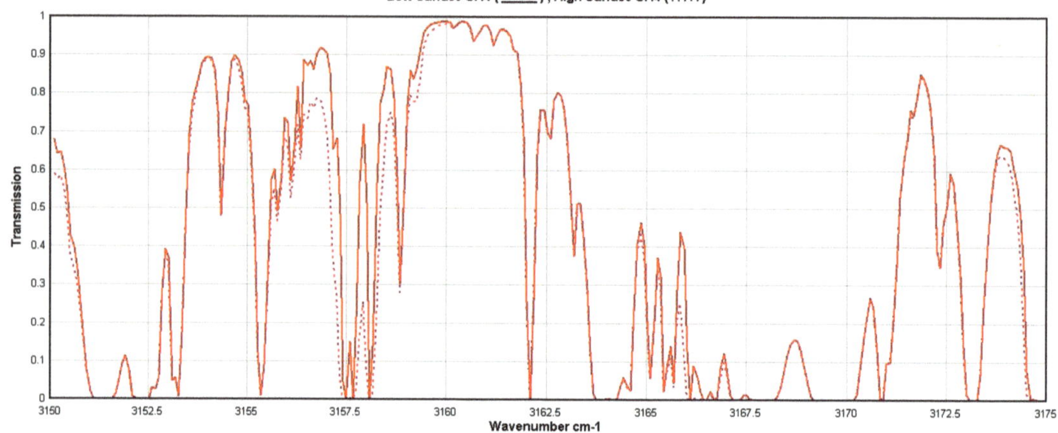

Methane Spectral Key
Atmospheric Transmission at 3150cm-1 (65deg slant angle, full atm.)
Low Surface CH4 (_____) , High Surface CH4 (. . . .)
Transmission
1
0.9
0.8
0.7
0.6
0.5
0.4
0.3
0.2
0.1
0
3150
3152.5
3155
3157.5
3160
3162.5
3165
3167.5
3170
3172.5
3175
Wavenumber cm-1

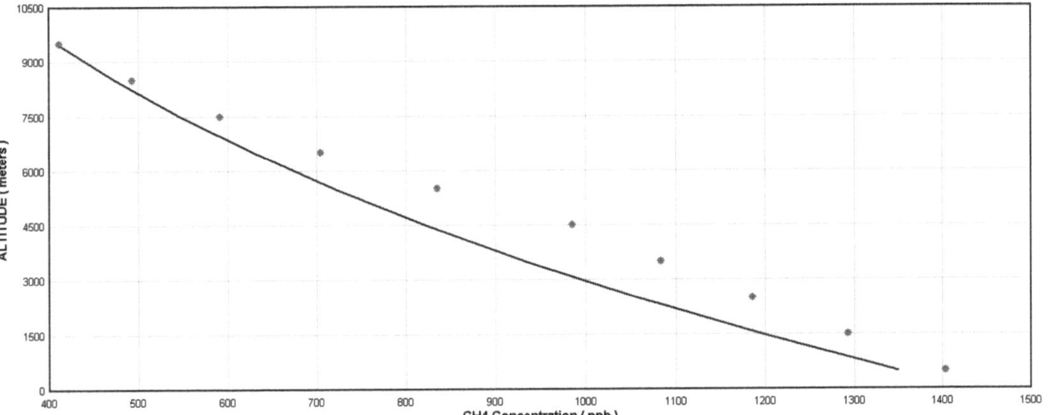

CH4 RETRIEVAL TEST #3
Test Profile #3 (ˆ) , Reference Mid-Latitude Summer CH4 Profile (___)
10 Layers , 45.DEG Look Angle (from zenith)
ALTITUDE (meters)
CH4 Concentration (ppb)

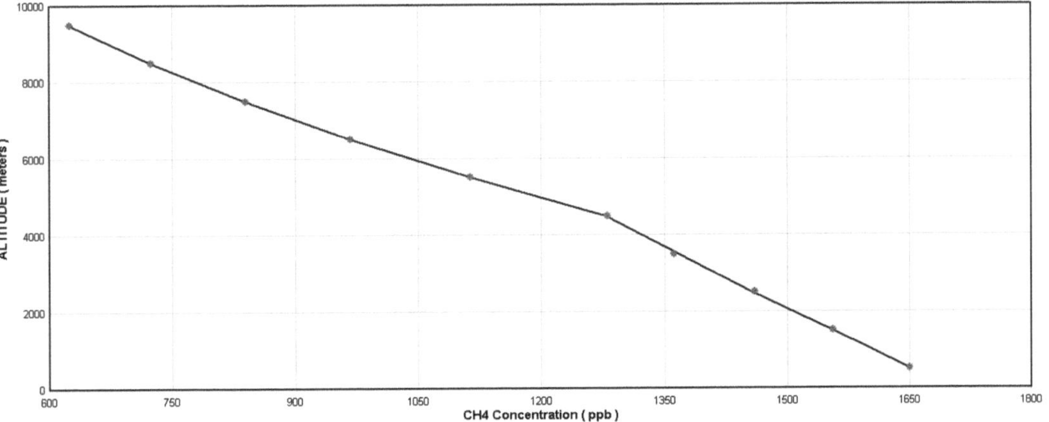

CH4 RETRIEVAL TEST #3
Retrieved Profile from spectrum (ˆ) , Actual Test#3 CH4 Profile(___)
10 Layers , 45.deg Look Angle (from Zenith)
ALTITUDE (meters)
CH4 Concentration (ppb)

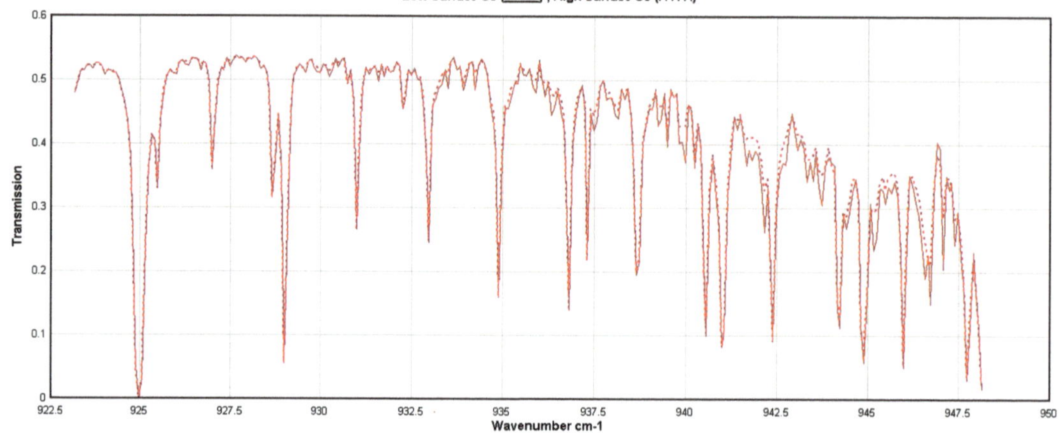

Ozone Spectral Key
Atmospheric Transmission at 923cm-1 (65deg slant angle, full atm.)
Low Surface O3 (_____) , High Surface O3 (. . . . .)
0.6
0.5
0.4
0.3
0.2
0.1
0
Transmission
922.5
925
927.5
930
932.5
935
937.5
940
942.5
945
947.5
950
Wavenumber cm-1

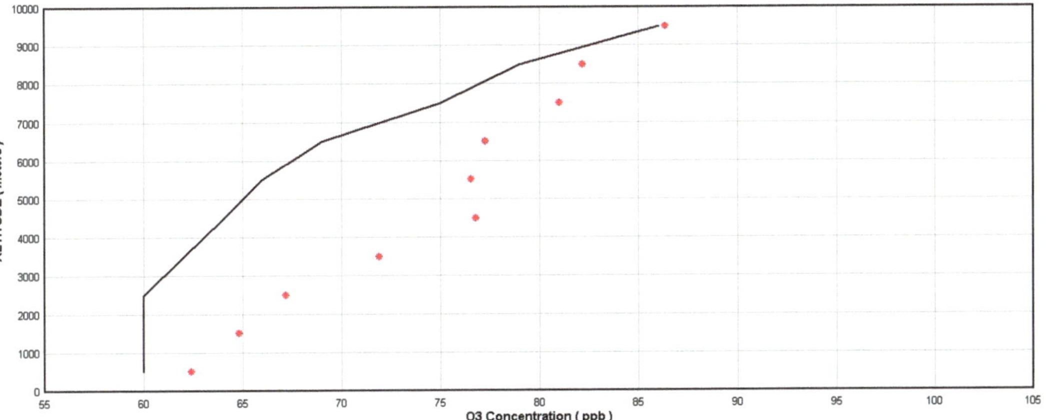

O3 TEST #3, 10 Layers, 65deg Look Angle
Test #3 O3 Concentrations (data points *)
Reference Mid-Lat O3 Concentration (solid line _____)
ALTITUDE (meters)
O3 Concentration (ppb)

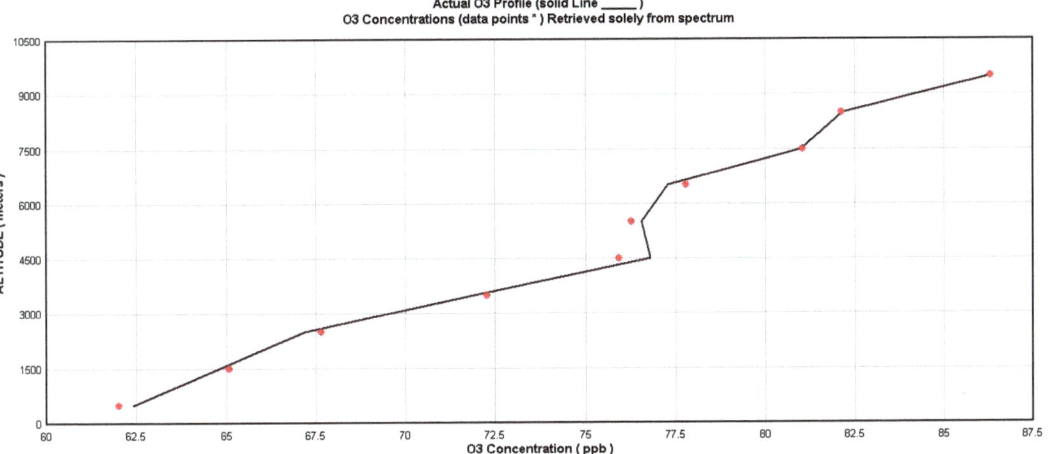

O3 PROFILE #3, 10 Layers, 65deg Look Angle
Actual O3 Profile (solid Line _____)
O3 Concentrations (data points *) Retrieved solely from spectrum
ALTITUDE (meters)
O3 Concentration (ppb)

www.ingramcontent.com/pod-product-compliance
Lightning Source LLC
Chambersburg PA
CBHW040251240726
48664CB00001B/349